Looking At The Cat

An Eye on Evolution

MARY JO NICKUM

Aquitaine Ltd
Phoenix, Arizona

"It is not the strongest species that survives, nor the most intelligent that survives. It is the one that is most adaptable to change."

Charles Darwin

Looking At The Cat

An Eye on Evolution

MARY JO NICKUM

Aquitaine Ltd
Phoenix, Arizona

Cover design by JD Smith Design

Library of Congress Control Number: 2016944892

ISBN-10: 0-692-67144-7
ISBN-13: 978-0-692-67144-3

www.aquitaineltd.com

Advanced Praise for
LOOKING AT THE CAT; AN EYE ON EVOLUTION…

"…you cover the subject very well…the book should be useful to students. We certainly need more books like this."

—Harley Shaw, Director The Cougar Network
Author of *"Soul among Lions, the Cougar as Peaceful Adversary"*

"Quite a story, well researched, well written and very courageous."

—Dr. Valerius Geist, wildlife researcher and
Professor Emeritus The University of Calgary, Alberta.

"I believe it would be great to have the information "out there" for students to read. Your LOOKING AT THE CAT; AN EYE ON EVOLUTION helps to fill that niche."

—Dr. Gene Ulrich National Teacher of the Year Finalist

DEDICATION

This book is dedicated to young people everywhere who
Don't like to or find it difficult to read and
Who, therefore, live on the fringes of a happy, healthy life.
Learn to enjoy reading and the world can be yours.

Believe in yourselves.

Why a Reluctant Reader Series?

Not all children read at the same level nor do all children enjoy reading. While a good many children do become excited and engaged in reading, especially in the primary grades, some are reluctant and disinterested. While a child may not show a natural interest in reading, this does not mean he/she cannot become a skilled and even enthusiastic reader. If the child has reached middle grade and is still disinterested, it's time to take action.

While any child, young or old, male or female can be a reluctant reader the largest number of unenthusiastic readers are adolescent boys. Research shows that a good number of boys who were avid readers in the elementary grades become disinterested in reading during their middle school years. However, there are a number of factors that may contribute to this shift—increasing complexity of material, peer pressure—one of the primary reasons seems to be they fail to see the connection between reading and "real" life.

School assignments, such as book reports, can become stressful for these students as well as parents and teachers. These children need material that is especially prepared to be relevant to the curriculum as well as to life. This material must be written to engage the reluctant reader, using images as well as text and presented in electronic and paper to fit classroom and leisure reading.

Books for the reluctant reader must be:

- Relevant to the curriculum (Common Core),
- Factual but engaging,
- Written to pique and hold interest,
- Presented in text and images,
- Available as ebook and in print, and
- Priced to be affordable for the individual student as well as bulk-priced to be attractive to schools.

Aquitaine, Ltd.'s Reluctant Reader Series, geared to middle grade readers, fills all these requirements. Our books have been thoroughly researched and edited by leading scientists and written by educators and librarians.

Illustration List

Figure 1a Prehistoric Forest 2
Figure 1b Prehistoric Forest 2
Figure 2 The Dirk-Tooth Cat Skull 3
Figure 3 The Simitar-Tooth Cat 4
Figure 4 Temperature Change 5
Figure 5 Protoceratids 6
Figure 6 Prehistoric Wild Horses 7
Figure 7a. First wave of Feline Migration 8
Figure 7b. Second wave of Feline Migration 9
Figure 8 Land Cracking From Earthquake 10
Figure 9 Continental Drift Map 10
Figure 10 Charles Darwin 11
Figure 11 Heritable Charges Over Time 12
Figure 12 Creodon 13
Figure 13 Nimravidae 15
Figure 14 Dinictis 16
Figure 15 Parallel Evolution 17
Figure 16 Miacidae 18
Figure 17 Viverravidae 19
Figure 18 *Proailurus Lemanensis* (first cat) 20
Figure 19 *Pseudaelurus* lorteti 21
Figure 20 Sabre-Tooth Cat 22
Figure 21 Modern cat *(Felis silvestris)* 23
Figure 22 Sabre-Tooth Cat Skull 24
Figure 23 *Dinaelurus crassus* 25
Figure 24 Simitar Cat Fossil 26
Figure 25 Rancho La Brea 28

Figure 26a Oil bubble in Tar Pit in downtown Los Angeles 28
Figure 26b Map of Los Angeles showing tar pit location 29
Figure 27 Pool of Water Covering Tar Pit 30
Figure 28 Friesenhahn Cave 32
Figure 29 American Scimitar Cat Skull 33
Figure 30 *Smilodon* Skeleton 34
Figure 31 Panthera leo atrox (American Lion) 35
Figure 32 A cave to where scimitar cats drug their prey to eat 36
Figure 33 Richmond Hill Fossil Site 38
Figure 34 Chadron Formation and extends from Nebraska to Saskatchewan, Canada 39
Figure 35 Sediment Layers Record Times 40
Figure 36 Forest to grassland from the late Oligocene through the Miocene (from about 30 million to 5 million years ago) 44
Figure 37 Sabre-Tooth Cat Skeleton 45
Figure 38 Martelli's Wildcat 49
Figure 39 European Wildcat 50
Figure 40 The Lynx 51
Figure 41 The Bobcat 52
Figure 42 Puma or Cougar 53
Figure 43 African Lion Pride 54
Figure 44 Cave Art of a Cat 58
Figure 45 Vogelherd Caves In Germany 59
Figure 46 European Lion 60
Figure 47 African Wildcat 61
Figure 48 Climate Change Over 5 Million Years 63

Foreword

Dear Educators and Parents,

I know many of you wonder why some people, especially children love to read and others never open a book unless it is a requirement. This dichotomy has bewildered educators and parents since literacy became a focus in our lives. Is it the topic, non-fiction versus fiction, the genre, the reading level, learning disabilities, lack of early exposure to reading in the home, or just no patience to sit down and read words on a page? Choose any one or more of these reasons and you still will not solve the problem.

We live in a world of instant visual gratification, that moves at a fast pace through television, videos, computers and on-line games. Getting children engaged in reading is a challenge that requires thinking outside the box. A book must catch the interest of the child and make him or her want to turn the next page. Simple to use in the classroom and meeting the core curriculum standards, *Looking at the Cat, An Eye on Evolution*, the first book in" The Reluctant Reader Series" by Mary Nickum provides a curriculum vehicle that teachers and parents can use to encourage teens to want to read about the history of cats going back 8,000 years. The book incorporates diagrams and photos for visual arts, language arts, biology, history, geography, and archeology. It satisfies learning on all levels, appealing to the most reluctant reader, especially if they relate to cats. It is this connection with the contents, storyline and presentation of a book that engages children and captures their interest in wanting to read more.

Fran Orenstein, Ed.D.
Educator, Counselor and Author

(**www.franorenstein.com**)

Table of Contents

Illustration List	ix
Foreword	xi
Table of Contents	xiii
Chapter One The Cat	1
Chapter Two What is Evolution?	11
Chapter Three How Did We Find Out About the Old Cats?	23
Chapter Four Why Did the Cats Change?	35
Chapter Five When Did the Modern Cats Emerge?	38
Chapter Six When Did People Start Associating with Cats?	46
Chapter Seven Will the New Cats Change and Become Old Cats?	51
Glossary	54
Facts of Evolution	58
Appendix I Status and Population Trends for Endangered Cats	60
Appendix II Wild Cat List by Continent	61
More Books about Cats	63
Illustration Credits	65
Resources	66

Felis catus, also called the domestic cat or house cat, is a small feline carnivorous mammal commonly used as a pet. Its most immediate pre-domestication ancestor is believed to be the African wild cat, *Felis silvestris lybica*. The cat has been living in close association with humans for at least 3,500 years; the Ancient Egyptians routinely used cats to keep mice and other rodents (mostly rats) away from their grain. The history of the domestic cat may stretch back even further, as 8,000-year-old bones of humans and cats were found buried together on the island of Cyprus.

Chapter One

The Cat

There are 64 million domestic cats in America. Although domestic cats are not native to North America, they are welcome in our homes. The total number of cats in the world is not recorded, but some say the total count may be as many as 600 million. All house cats have one common ancestor, the European wildcat. The ancestors of the European wildcat can be traced to the Miocene epoch, 25 million years ago. This is how it all began.

Imagine what the world was like 25 million years ago. Forget what you see today. Things were different then, no roads, no cars, and no buildings, nothing that is familiar. There were no humans. Even the plants and animals were different (Figure 1a and Figure 1b).

Figure 1a Prehistoric Forest

Figure 1b Prehistoric Forest

The **Miocene** or "less recent" is so called, because it contains fewer modern animals than the next, **Pliocene, epoch**. The Miocene, which lasted 18 million years, was a time of huge transition, the end of the old prehistoric world and the birth of a more recent sort of world. It was also the high point of the age of mammals and cats are mammals.

The swift intelligent cat- and dog-type carnivores replaced the slow, clumsy creodonts as the dominant predators. **Saber-tooth cats** roamed the planet. The **dirk-tooth cats** (Figure **2)** had two long, narrow upper canine teeth and short legs. They looked like bears.

Figure 2 The Dirk-Tooth Cat Skull

Figure 3 The Simitar-Tooth Cat

The **scimitar-tooth cats** (Figure **3** had two shorter, broader upper canines and long legs. They were built like a cheetah.

These cats were well suited for hunting in the grassland ecosystems that gradually replaced the forests during the Miocene. In Eurasia and North America, the spread of grasslands forced an evolutionary change in **herbivorous** mammals, with the old forest browsers giving way to the new prairie grazers. Gone were the forests of **lepidodendron, cordaites, calamites** and ***Sigillaria***. Mammal diversity reached its peak during the Miocene. Numerous species began then became **extinct** in response to the changing environment (Figure 4).

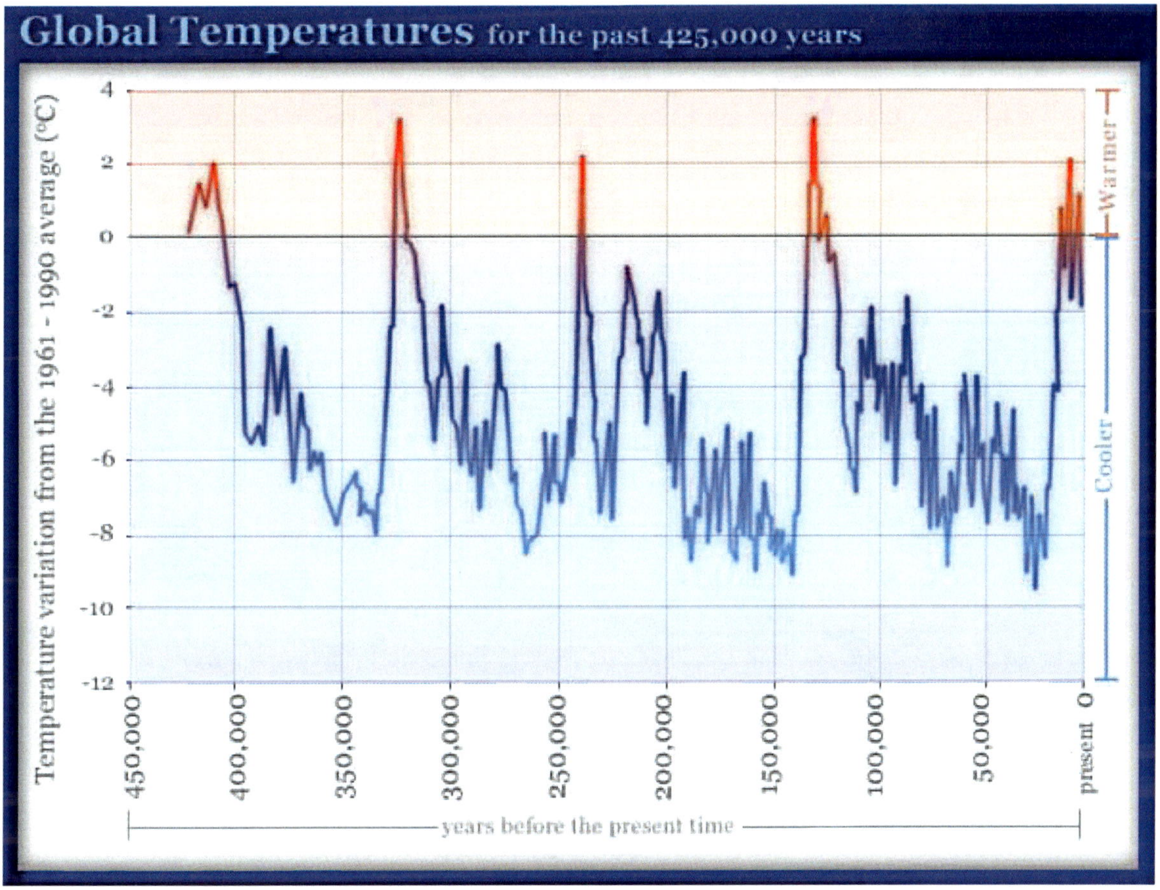

Figure 4 Temperature Change

Many of the mammals that flourished in the Miocene were hoofed grazers or browsers. This epoch was marked by further evolution of horses, which became plains type animals as large as ponies. The chalcitheres, camels, rhinoceroses, and anthropoid apes inhabited Southern Europe, Asia, and Africa, all prey-food for the cats. In addition, this period saw the appearance of the mastodons, raccoons, and weasels. The first deer and giraffes appeared, along with the first hyenas, which competed with the cats for these food items (Figure 5).

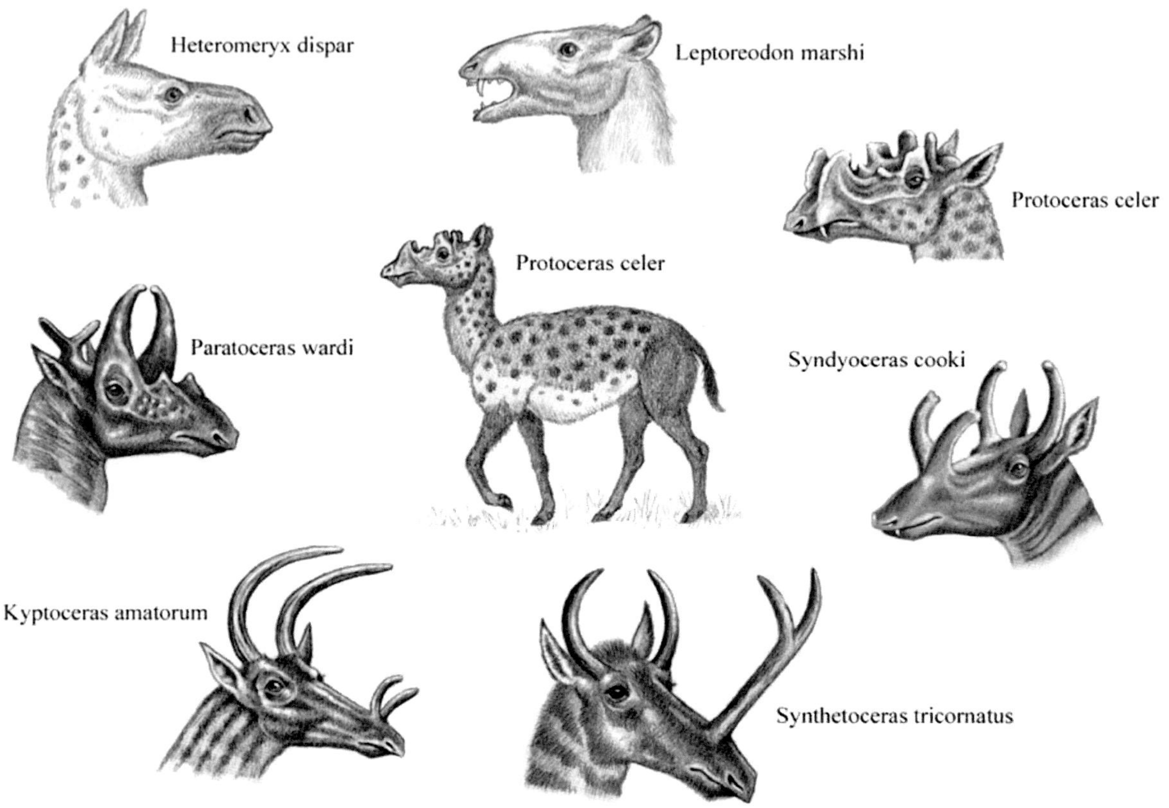

Figure 5 Protoceratids

Great migrations took place during the mid-Miocene. Mastodontids spread out from Africa to Eurasia and North America. Cats, giraffes, pigs, and cattle went the opposite direction, from Eurasia to Africa. Horses and camels spread out from North America into Eurasia. In North America, species of rhino roamed the countryside alongside as many as a dozen horse species (Figure 6).

Figure 6 Prehistoric Wild Horses

At different times over the next 15 million years, the ancestors of modern cats spread from Asia to other parts of the world. Ancient changes in sea level created temporary land links that allowed cats to cross between continents and regions now separated by water. In other cases, sea levels that were as high, or higher, than today separated groups of cats. Both migration and separation affected cat evolution. Most notably, some types of cats that evolved in the Americas from Asian ancestors later migrated back into Asia. Examples include the cheetah, which spread from North America to Asia, and then to Africa; and, the lynx, which spread from North America into Asia and Europe (Figure 7).

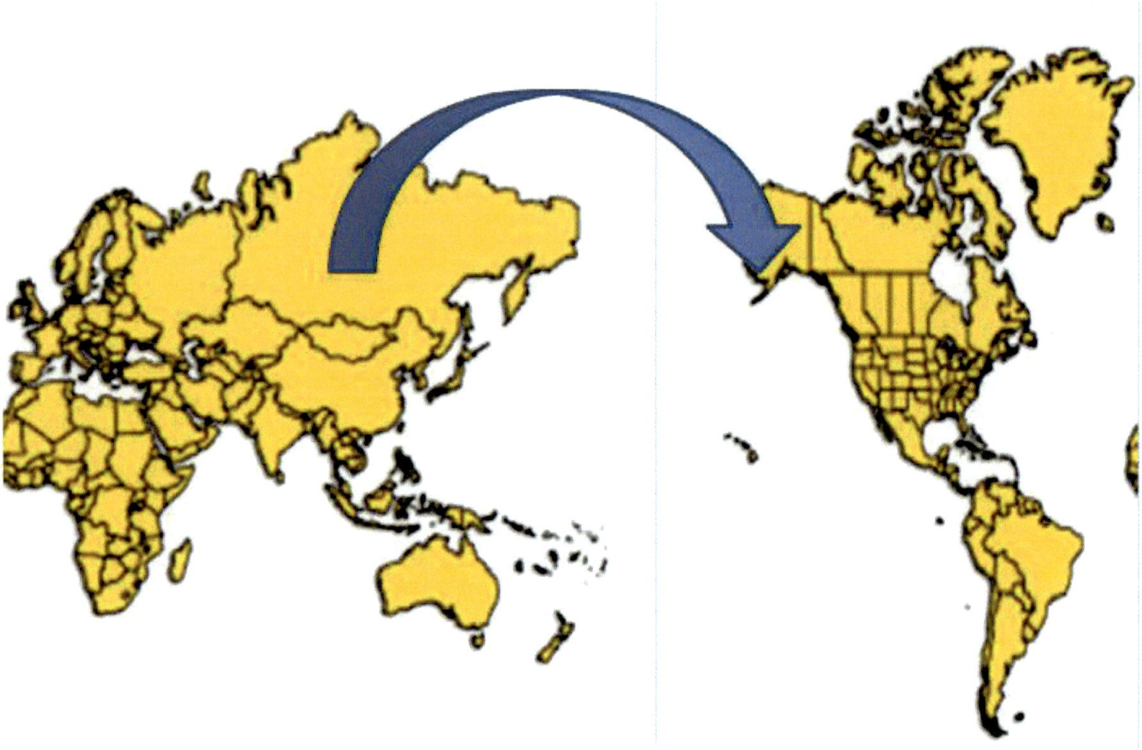

7a. First wave of Feline Migration

7b. Second wave of Feline Migration

During the Miocene, the earth was active; volcanoes erupted, creating mountains that gradually wore away. Dry lands became deep seas that eventually dried, leaving thousands of feet of many-colored **sediments**. Wind and water erosion etched out the softer areas and sometimes sculpted towering formations. Earthquakes shook the land, buckling and cracking the surface (Figure 8).

Figure 8 Land Cracking From Earthquake

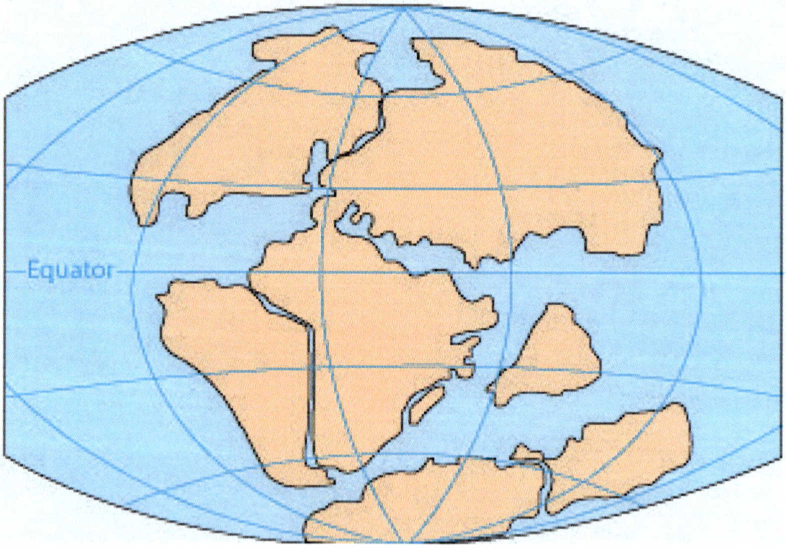

Figure 9 Continental Drift Map

Africa became more arid during this epoch and, as India continued to drift, it collided with the Asian mainland and formed the Himalayas. As South America moved north, the passageway between the continent and Antarctica opened (Fig. 9). The resulting circulation of cold waters around Antarctica led to the formation of deep, cold bottom waters in Earth's oceans. All of this contributed climate changes favoring the **evolution** of the cat.

Chapter Two

What is Evolution?

Figure 10 Charles Darwin

Charles Darwin (Figure 10), a scientist in the 1800s, introduced the basic concepts of evolution. Most people in his time believed, and some still do, that every kind of animal on earth has appeared here through an act of special creation. Darwin said this is not true. All forms of life have evolved out of other forms that lived at an earlier time, and those, in turn, had evolved out of still earlier forms back into the distant past. According to Darwin's law, evolution is a process that results in heritable changes in a population spread over many generations (Figure 11).

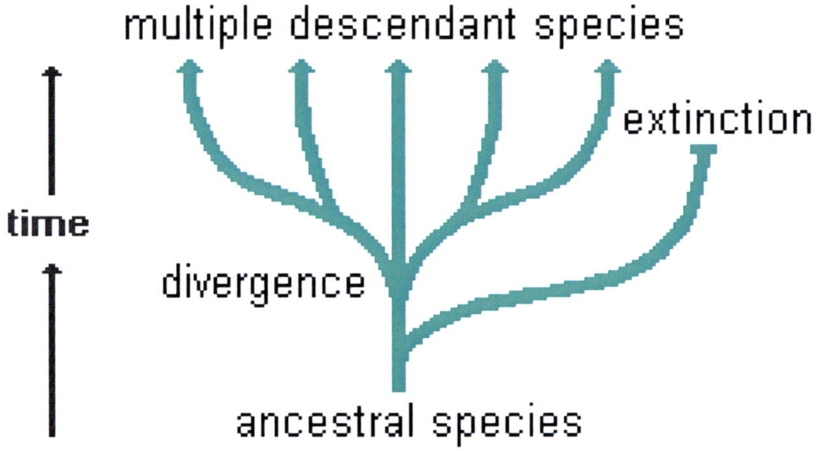

Figure 11 Heritable Charges Over Time

Darwin's theory consisted of three major ideas:

- All living things—two million species—developed from a few primitive, simple organisms. Over millions of years, gradual changes occurred in these first species.

- After studying reports that populations multiply faster than their available food supply, Darwin concluded that nature acts as a selective force in which the least able to live and reproduce die and the fit survive.

- Darwin proposed that individuals of species that have adapted successfully live long enough to reproduce. The offspring of the survivors are better suited to their changing environment and, they too, will survive.

Biological evolution refers to gradual change in populations and not to sudden change in individuals and those changes must be passed on to the next generation. For evolution to progress, populations of living things develop **adaptations** to their environment. A population grows until it reaches the limit of its resources. In the struggle for existence that results, individuals with traits that help them to overcome adverse changes in the environment, everything from climate to disease to attacks from predators, are more likely to survive and produce offspring. The offspring inherit the favorable traits from their parents and carry them on to the next generation.

On the other hand, individuals with traits that handicap them in the struggle are less likely to reach maturity and reproduce, causing those traits to disappear from the population over time. Remember this is a slow process. You probably won't see any evolutionary changes in your lifetime. Let's look at creodonts to see how this happened to make a cat.

Figure 12 Creodon

This is not as simple as it sounds. Creodonts (Fig. 12), though meat-eaters, were very different from current day carnivores. It is not known exactly why the creodonts were replaced by the Carnivora. It may be because of their smaller brains and their locomotion, which was somewhat less energy-efficient (especially while running).

Their limb structure limited leg movement to a vertical plane, as in horses; they were unable to turn their wrists and forearms inward to trip, slash, or grab prey as modern carnivores can. Creodonts had to depend entirely on their jaws to capture prey, which may be why creodonts generally had a larger head size in relation to their bodies than carnivores of similar stature. The creodont lumbosacral spine, lower part of the spine toward the hips and back legs, was not arranged as efficiently for running as in Carnivora. The arrangement of the teeth was also somewhat different. In the miacids and as with the modern Carnivora) the last upper premolar and the first lower molar are the carnassials, allowing grinding teeth to be retained behind for feeding on non-meat foods (the Canidae are the closest modern analog to miacid dentition). Miacids were small carnivores, superficially marten- or civet-like with long, little bodies and long tails. Some species lived mostly in trees, while others lived on the ground. In creodonts, the carnassials were further back—either the first upper and second lower molars, or the second upper and third lower molars. This committed them to eating meat almost exclusively. These limits may have created important disadvantages over millions of years.

There were two forms that we now call families, one cat-like and the other dog- or hyena-like. The cat-like group or **genera** came before the saber-toothed cat forms. Although there were a number of cat-like animals, they differed enough not to be considered cats as we know them. True cats were not present before about 30 million years ago, in the mid to late Oligocene, the fourth epoch. The earlier cat-like animals were called nimravids (Figure 13).

Figure 13 Nimravidae

They showed several features that we associate with cats, including long legs ending in feet with claws, a short face with slicing teeth and sharp, pointed canines. They are sometimes called False Saber-Tooths, an extinct family of fierce carnivores. While the nimravids resembled the true saber-toothed cats, they were not closely related.

Figure 14 Dinictis

Dinictis (Figure 14) was a member of the nimravid family. Dinictis walked flat-footed, unlike modern cats. It lived in the plains of North America around 40 million years ago, from the late Eocene and early Oligocene. The Eocene and Oligocene, the second and third epochs of the Tertiary period, marked the emergence of the first modern mammals.

The cat family, now called Felidae, emerged during the late Oligocene alongside another group, the successful Nimravidae. This is an example of parallel evolution (Figure15).

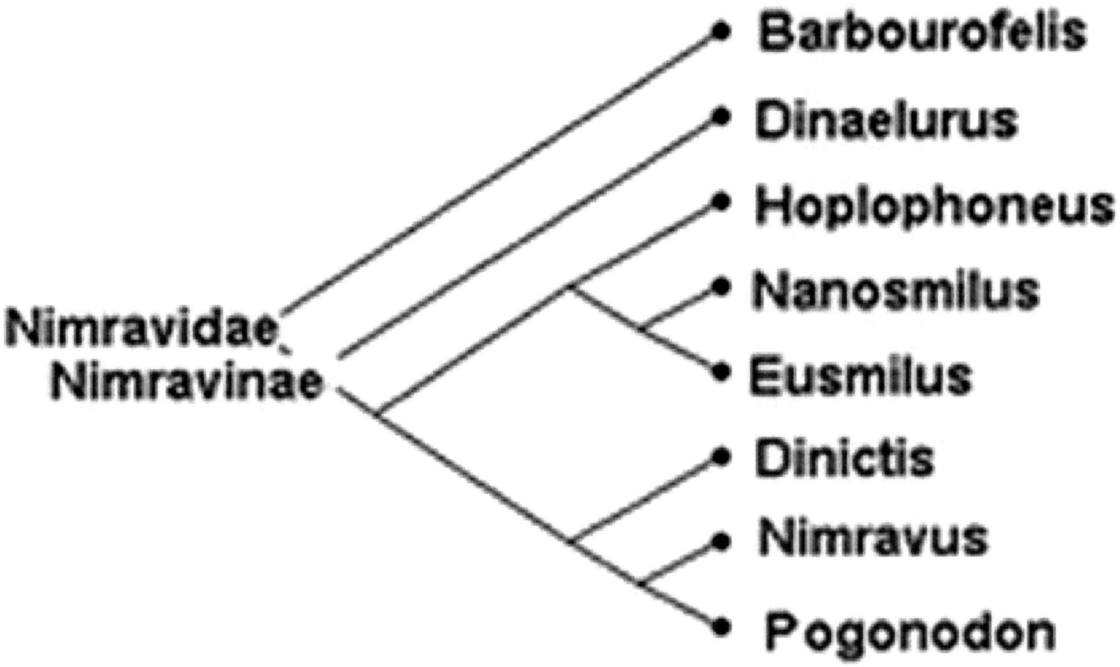

Figure 15 Parallel Evolution

Carnivorous mammals evolved from Miacidae, small pine marten-like insectivores (Figure 16), which lived 60-55 million years ago. The miacids

Figure 16 Miacidae

split into two lines: Miacidae and Viverravidae. Miacidae gave rise to Arctoidea/Canoidea group (bears and dogs) while Viverravidae (Figure 17)

Figure 17 Viverravidae

gave rise to Aeluroidea/Feloidea group (cats, hyenas, civets, mongooses) around 48 million years ago. The Viverravidae also gave rise to a group called Nimravidae. The Nimravids were cat-like creatures, evolving in parallel with true cats. They are not part of true cat lineage and have left no living descendants.

The first true cat to arise from Viverravidae was Proailurus ("first cat", Figure 18) around 30 million years ago. The best-known species was *Proailurus lemanensis*, found in France. *Proailurus* was a small weasel-like cat with relatively short legs and a long body. It had one more premolar on each side of its bottom jaw than do modern cats.

Figure 18 *Proailurus Lemanensis* (first cat)

About 20 million years ago, *Proailurus* gave rise to *Pseudaelurus*. *Pseudaelurus* were Miocene ancestors of cats. *Pseudaelurus lorteti* (Figure 19) was about the size of a large lynx, while *P. validus* was the size of a large lynx or small puma.

Figure 19 ***Pseudaelurus*** **lorteti**

Three other species of early cat are described as *Pratifelis, Vishnufelis* and *Sivaelurus* (*Sivaelurus chinjienis*). *Pseudaelurus* went on to split into two major groups: the Machairodontinae (true sabre-tooths, Figure 20) and *schizailurus* (the ancestor of the modern day Felidae group). *P. schizailurus* was almost indistinguishable from the modern cat.

Figure 20 Sabre-Tooth Cat

Although animals with feline traits appeared as early as 30 million years ago, ancestors of modern cats were not evident until the mid-Miocene period. About 20 million years ago, during the late Miocene, the cats split into to two distinct groups—one that led to the modern cats (Figure 21).

Figure 21 Modern cat ***(Felis silvestris)***

and another that led to the now extinct saber-toothed cats (Figure 22).

Figure 22 Sabre-Tooth Cat Skull

In contrast to the large-toothed forms, a more "normal" catlike species known as *Dinaelurus crassus* (Figure 23) emerged. It looked somewhat like a cheetah.

Figure 23 ***Dinaelurus crassus***

Chapter Three

How Did We Find Out About the Old Cats?

The weight of the evidence should be proportioned to the strangeness of the facts.

---Pierre-Simon Laplace (1749–1827)

Figure 24 Simitar Cat Fossil

Fossils (Figure 24) tell us everything we know about the old cats. But aren't fossils just bones or imprints? How can we tell anything from bones? How do we know how old they are? Fossils are the remains of a plant or animal of a past age preserved in earth or rock. The remains are sealed off quickly from air and moisture after death. Fossils are formed as minerals that are harder than the surrounding sediments; other fossils are formed in coal deposits over millions of years.

Bone is living tissue and, though it doesn't look like it, it does bleed. Being mostly protein, it will begin to decay after death and will disappear over time in acidic conditions. Thus, no fossils will be found in those areas. Of all things, rapid burial is one of the most effective means of making sure that remains will be preserved. There are several places in the world, some here in the US, which are particularly good for finding fossils.

Caves are considered the best, especially limestone caves. Even when the deposit has become compressed or cemented, the bones can often be recovered by etching away the hard material with a weak acid solution. The bone will be in excellent condition and can be studied like fresh bone.

Suitable conditions can occur in areas where the bones have been covered by water-laid sediments, such as riverbeds or lakeshores. One of the most impressive sets of fossils in North America was found in the famous Rancho La Brea (Figure 25).

Figure 25 Rancho La Brea

in California, in the middle of downtown Los Angeles (Figure 26a &b).

Figure 26a Oil bubble in Tar Pit in downtown Los Angeles

Figure 26b Map of Los Angeles showing tar pit location

Saber-toothed cats, lions, dogs and their prey became trapped in the sticky material, lurking under pools of water (Figure 27).

Figure 27 Pool of Water Covering Tar Pit

Tar pits form when crude oil seeps to the surface through cracks in the Earth's crust; the light fraction of the oil, the part of crude oil that dissolves in water, evaporates, leaving behind the heavy tar, or asphalt, in sticky pools.

Le Brea Tar Pits

In 1781, José Longinos Martinez described the area: "…there are nearly 20 springs of liquid petroleum, pitch, etc…in the middle of a great plain of more than 20 leagues in circumference, there is a great lake of pitch, with many pools in which bubbles or blisters are continually forming and exploding…In hot weather animals have been seen to sink in it and when they tried to escape, they were unable to do so because they were stuck and the lake swallowed them."

Tar from the La Brea tar pits was used for thousands of years by local Native Americans, as a glue and as waterproof caulking for baskets and canoes. After the arrival of the Spanish, the tar from these pits was used for roofing by the inhabitants of the nearby town of Pueblo de Nuestra Señora la Reina de Los Angeles.

Today, the Page Museum is located at the Rancho La Brea Tar Pits in the heart of Los Angeles. It contains the only skull of a saber-toothed cat known to be found with its jaw closed. This specimen not only shows how their teeth functioned in the closed position but also shows that a saber-tooth cat could fully close its jaw while alive. Through windows at the Page Museum Laboratory, you can watch bones being cleaned and repaired. Outside the Museum, in Hancock Park, life-size replicas of several extinct mammals are featured. The museum's rich displays of extinct creatures and prehistoric landscapes gives a detailed picture of life in North America from 40,000 to 4,000 years ago.

About 8-12 gallons a day of tar still ooze and bubble to the surface, occasionally trapping insects and small animals.

Another important US site is Friesenhahn Cave (Figure 28) near

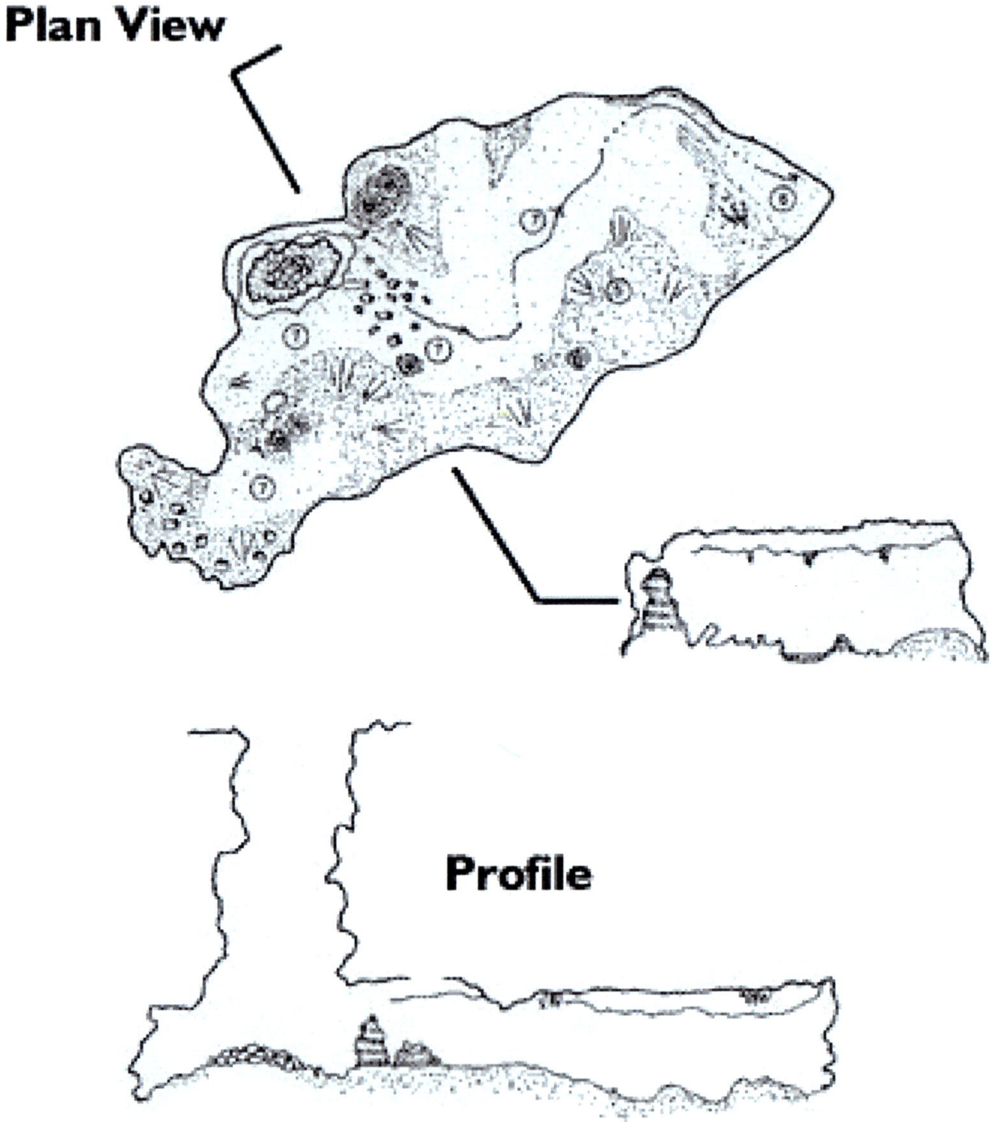

Figure 28 Friesenhahn Cave

Friesenhahn Cave

Present-day entrance to the cave is gained by descending a ladder 45 feet below the surrounding land surface. The original cave opening has been closed because of collapsed limestone blocks and rock debris that washed in from the surface. Sometime later, the new vertical shaft (sinkhole) opening was formed, once again providing a natural, but straight vertical decent into the cave.

The present dimensions of the cave are approximately 60 feet long and 30 feet wide. The rather modest dimensions of the Friesenhahn Cave belie its paleontological importance. It was the first cave in Texas studied by the scientific community for fossils. It is recognized now as one of the richest and most important Pleistocene fossil sites in North America.

San Antonio, Texas. Excavations in the cave fill have yielded bones and teeth of more than 30 genera of mammals, reptiles, amphibians, and birds. One of the most important finds was a nearly complete set of remains of an adult American scimitar cat (Figure 29, *Homotherium* serum) and a nearly complete infant cat skeleton.

Figure 29 American Scimitar Cat Skull

This cat had slender-legs, with long forelegs, short, powerful hind legs and a short tail. Its long upper fangs were like curving steak knives. Although much rarer than the saber-tooth cat (Figure 30, *Smilodon* fatalis)

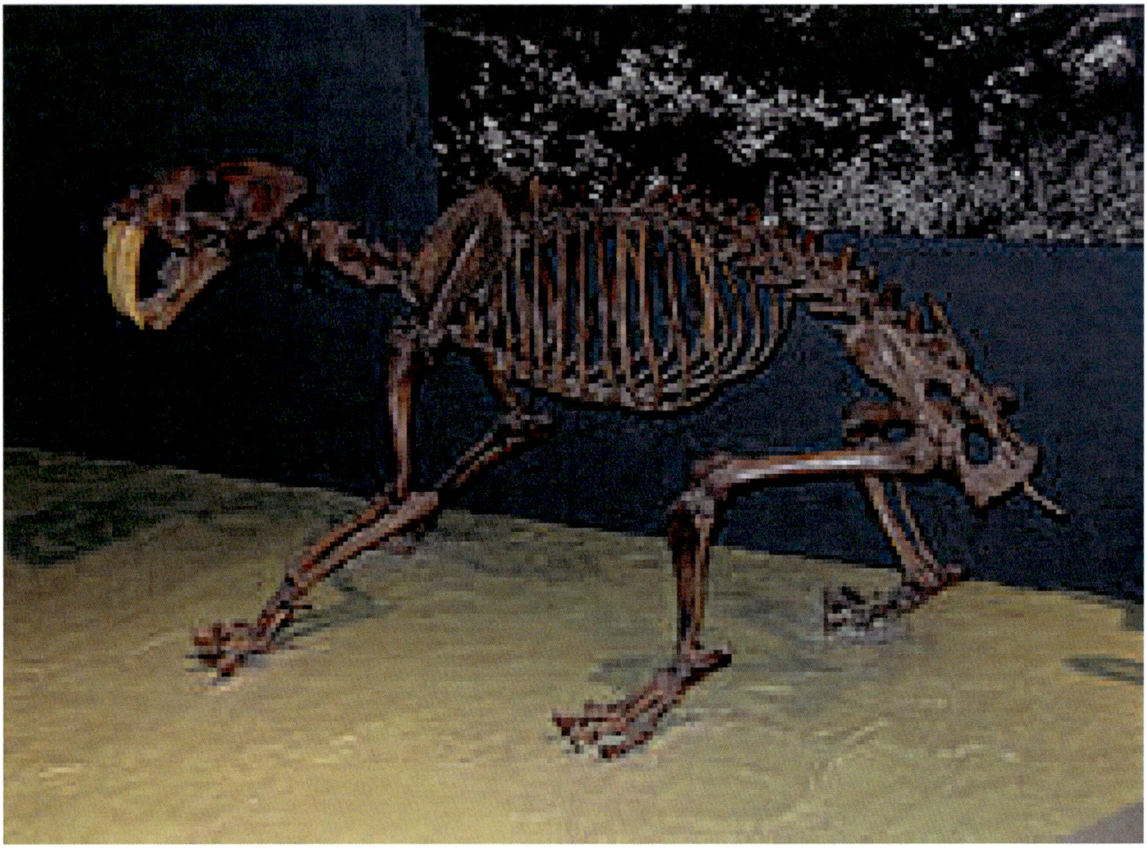

Figure 30 *Smilodon* Skeleton

and the American lion (Figure 31, *Panthera leo atrox*), it seems to have

Figure 31 ***Panthera leo atrox*** **(American Lion)**

been the only saber-tooth cat to have lived in Eastern Beringia, modern Alaska. An exceptional number of young mammoth (*Mammuthus columbi*) remains were in the cave also. The bones were scattered, some appear to have pits or sheared surfaces, strongly suggesting that scimitar cats hunted young mammoths and drug them into the cave to eat (Figure 32).

Figure 32 A cave to where scimitar cats drug their prey to eat

An earlier form of a nimravid, a saber-toothed cat, is known from a Richmond Hill fossil site (Figure 33) in the northern Black Hills, South Dakota.

The Chadron Formation

The Brule Formation, a later part of the Chadron Formation, is the uppermost formation in the White River Group. This important fossil formation was named for the Brule Indians that lived in western Nebraska and South Dakota. The Brule Formation is exposed over a vast area of the West, including Nebraska, North and South Dakota, Wyoming, and Colorado, and yields abundant fossils as layers are eroded. It is composed of pinkish/brown to gray/green interbedded claystones, calcareous mudstones, siltstones, freshwater limestones, tuffaceous beds, and cross-bedded sandstones. The Brule Formation, often banded pink and white, makes up most of the Badlands cliffs in the Badlands National Park, South Dakota.

Toadstool Geologic Park is located in the Oglala National Grassland in far northwestern Nebraska. In the early Miocene, around 30 million years ago, a broad, shallow river flowed through this hot and humid area, attracting prehistoric mammals that came to drink. The currents carried and deposited layer upon layer of volcanic material, sand, and silt. Over time, these deposits were cemented together to form rocks containing the fossil record of early Great Plains animals that roamed the area 30 million years ago. As the Cretaceous sea receded, the High Plains emerged, gradually becoming the gently rolling grasslands of today.

The layers of earth and rock begin with the oldest Eocene-age Chadron Formation, followed by the Oligocene Brule Formation, and finally the Miocene age rocks of the Arikaree Group. Tracks are especially abundant in the Brule Formation, giving scientists insight into the behavior of the Oligocene mammals.

Figure 33 Richmond Hill Fossil Site

This site is part of the Chadron Formation and extends from Nebraska to Saskatchewan, Canada (Figure 34).

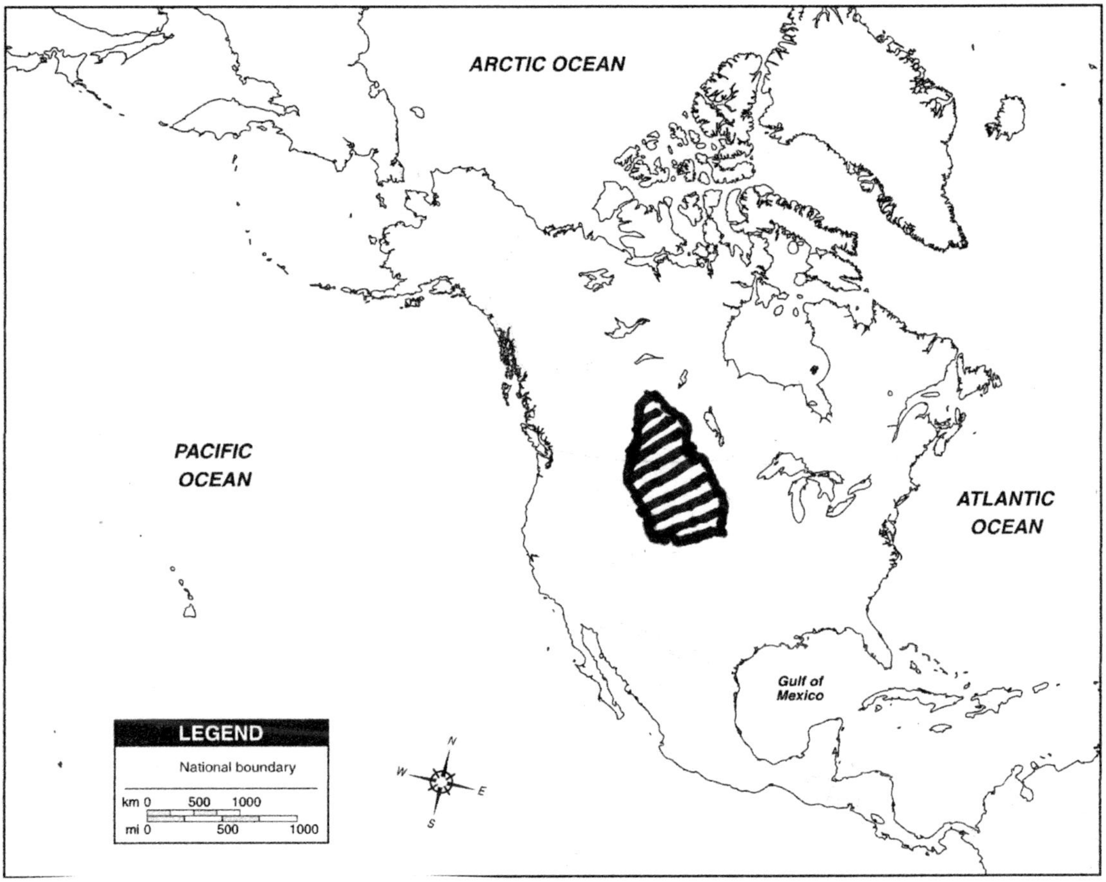

Figure 34 Chadron Formation and extends from Nebraska to Saskatchewan, Canada

All together, we have a great variety of Pleistocene vertebrate fossils here in the US. Several other sites around the world have yielded many complete specimens, including Cerro Batallones near Madrid, Spain, Senèze in France, Bolt's Farm in South Africa and Shansi in China. The keys to the fossil process are time and the right set of conditions.

All of the cats that we are talking about have lived during the past 30 million or so years. But, how do we know the length of time that has passed since they died? According to geologists who study deposits of sediment and rocks, the latest deposit, usually the ones closest to the (Figure 35)

Figure 35 Sediment Layers Record Times

surface, is the youngest and that group of fossils found there allows us to correlate that deposit with others elsewhere. This is called biostratigraphy and forms the basis of much of the geological timetable.

The absolute age of the deposit can now be established. A most useful technique involves measuring the amount of material produced by radioactive decay of materials encrusting the fossil. Absolute dates,

therefore, usually are attached to the deposits in which the fossils occur rather than the fossil itself. One exception to this is radiocarbon dating, which uses the ratio of radioactive carbon to nonradioactive carbon to assess the age since death of the animal or plant. Because the rate of decay is "rapid," it can be used only for fossil animals and plants that have died within the past forty thousand years, too short a time for our cats.

The fossil record of the small cats is poor because their ancestors mainly inhabited tropical forests where their carcasses were poorly preserved. We have more information about the bigger cats because they lived in open areas where conditions for fossil formation were better. From fossil remains of the various species of big cat, lion, tiger, leopard and jaguar, it has been possible to piece together an overview of the various relationships between today's cat species. But, we have not found all of the fossils. As new fossils are found, scientists may need to revise their conclusions. Fossils are a record of change.

The absolute age of the deposit can now be established. A most useful technique involves measuring the amount of material produced by radioactive decay of materials encrusting the fossil. Absolute dates, therefore, usually are attached to the deposits in which the fossils

occur rather than the fossil itself. One exception to this is radiocarbon dating, which uses the ratio of radioactive carbon to nonradioactive carbon to assess the age since death of the animal or plant. Because the rate of decay is "rapid," it can be used only for fossil animals and plants that have died within the past forty thousand years, too short a time for our cats.

The fossil record of the small cats is poor because their ancestors mainly inhabited tropical forests where their carcasses were poorly preserved. We have more information about the bigger cats because they lived in open areas where conditions for fossil formation were better. From fossil remains of the various species of big cat, lion, tiger, leopard and jaguar, it has been possible to piece together an overview of the various relationships between today's cat species. We have not found all of the fossils yet. As scientists find new fossils, they may need to revise their conclusions. Fossils are a record of change.

Chapter Four

Why Did the Cats Change?

As the plant life evolved with climatic change, so did the animal life. Noticeable change in climate began in the late Oligocene and continued into the Miocene. The drying of the landmasses caused forests to disappear and be replaced by grasslands (Figure 36).

Figure 36 Forest to grassland from the late Oligocene through the Miocene (from about 30 million to 5 million years ago)

Large grazing animals developed in these grasslands, providing prey for those meat-eating animals, such as cats that could stalk and kill prey, or hunt in groups and overwhelm an individual. This allowed them to be successful and, thus, survive–one important ingredient in evolution.

To be successful, a population must do more than survive. Individuals must reproduce to pass on to their young the traits that led to their success. Those who were not successful failed to reproduce and, therefore, their genetic line was lost. This describes the words that you have heard to summarize evolution, "survival of the fittest."

An example of a large cat that did not survive is the saber-tooth (Figure 37). Saber-tooths evolved independently at four different

Figure 37 Sabre-Tooth Cat Skeleton

times. There were saber-tooth creodonts, saber-tooth marsupials, saber-tooth nimravids and, at last, a saber-tooth **neofelid**.

The neofelid is the beginning of the cat that we recognize today. Scientists have been unable to solve the mystery of how saber-tooth cats were able to bite or eat. Many odd **hypotheses** have been proposed to explain the function of the long teeth, including their use as a can-opener on armadillos, a tool to grub for mollusks like walrus tusks, a way of stabbing with a closed mouth, and even as a tree climbing aid. Nevertheless, in the end, the saber-tooth cats were not successful.

Chapter Five

When Did the Modern Cats Emerge?

There are 41 known species of felines in the world today, all of which have descended from a common ancestor about 10.8 million years ago. Present-day cats are divided into two groups: the big cats of the genus *Panthera*, called roaring cats and the small cats, called purring cats, placed in the genus *Felis*.

Hunting Techniques of Cats

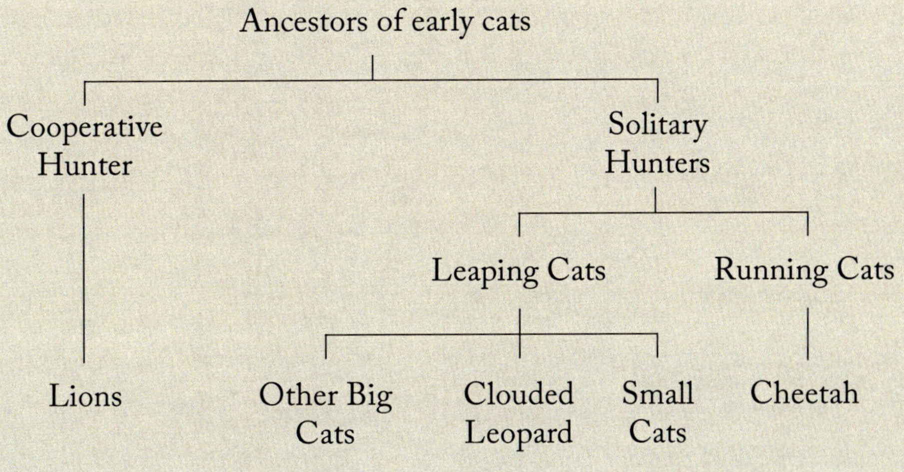

A few words about naming the animals, the generic category starts the process of organizing the names and the organisms. *Felis catus* is the house cat; *Felis lynx* is the lynx; *Felis concolor* is the North American mountain lion or cougar. The fact that all are named to the genus *Felis* indicates that zoologists think these cats are closely related, not closely enough to be *catus* or *concolor*, but more closely than to be members of *Panthera*, the genus for the lion. There are many levels of category beyond the genus. The next higher level after genus is family. The family Felidae includes *Felis* and *Panthera*, for example. And above family is order. The family Felidae is one of many families: Canidae (dogs), Usidae (bears), and so on in the order Carnivora, the meat eaters.

Until recently, zoologists often classified the smaller species of modern cats together in the single genus *Felis*, with the big cats usually kept separate as species of the genus *Panthera* and the cheetah in its own genus, *Acinonyx*. Much of this classification system is changing based on DNA work and other genetic studies that suggest a complex history that separates living members of the cat family into distinctlineages. Many cats, formerly classified as species of *Felis,* do not have the same common ancestors and so are not grouped together. Nonetheless, hybrids between different species of cats are often fertile and can produce fertile offspring.

The vocal distinction is important because it is a readily identifiable characteristic in all cats. In roaring cats, the hyoid bone in the throat is flexible while it is solid or ossified in the purring cats. Let's look at the roaring cats first.

Africa was the place for early lion development. Then, the cave lion evolved in Europe and the American lion probably evolved in eastern Asia and spread through North and Central America as far south as Peru in South America. The oldest known leopard remains are from India, but that was only about 2 million years ago. Fossil tigers were found in southeastern Asia. Those fossils date back to the late Pliocene/early Pleistocene.

The ancestor of the European wildcat, called Martelli's wildcat (Figure 38), is from the Early Pleistocene. By the end

Figure 38 Martelli's Wildcat

of the Pleistocene, surviving wildcats were smaller. Most scientists consider the European wild cat (Figure 39) as the ancestor of the

Figure 39 European Wildcat

This ranked system reflects scientists' opinions about the closeness of the relationships between various species and genera:

Kingdom: Animalia

Phylum: Chordata (Vertebrates)

Class: Mammalia

Order: Carnivora

Family: Felidae

Subfamily: Pantherinae
Uncia uncial
(snow leopard)

Genus:
Panthera; Felis; Acinonyx

Species:
Panthera leo (lion)
Panthera tigris (tiger)
Panthera onca (jaguar)
Lynx rufus (bobcat)
Felis lynx (lynx)
Felis concolor (cougar)
Felis catus (house cat)
Acinonyx jubatus (cheeta)

Figure 40 The Lynx

Its evolutionary center seems to be in Africa in the late Pliocene, about four million years ago. The lynx reached North America in the Pliocene, where it gave rise to the bobcat (Figure 41).

Figure 41 The Bobcat

In addition, pumas or cougars (Figure 42) are known to have been in

Figure 42 Puma or Cougar

North America from the Late Pliocene.

From the Miocene through the Pliocene into the Pleistocene, the climate changed from moist to dry, then repeated glaciations to a warming at the end of the Pleistocene. The changes in vegetation and, of course, habitat, triggered the evolutionary changes in the cat.

The old adaptations were no longer as useful. Through evolutionary adaptation to environmental change, marked differences came to distinguish one species from another. There were those built for speed over land or for agility in the treetops, those who hunted by day and others who were nocturnal in pursuit of prey. Those cats that were successful in the changing environment reproduced. Animals are said to adapt when, over many generations, a combination of successive, small, random changes in traits occur. These changes, called adaptations, make the animals best suited for their environment. The lion provides an example of adaptation.

The lion adapted to life on the open plains of Africa, the Serengeti, by forming prides (Figure 43). This communal, cooperative hunting lifestyle made them successful in this environment. Cooperative hunting is unique to the lion. No other cat hunts in this way. All other cats are loners, coming in contact willingly only for mating. The females of the pride often share responsibilities for caring for their young. They are the primary hunters, sharing their kill with the dominant male in "exchange" for the protection he provides to his pride.

Figure 43 African Lion Pride

Cooperative hunting is unique to the lion. No other cat hunts in this way. All other cats are loners, coming in contact willingly only for mating. The females of the pride often share responsibilities for caring for their young. They are the primary hunters, sharing their kill with the dominant male in "exchange" for the protection he provides to his pride.

Cats made dietary adaptations as well. Dental and other characteristics suggest that cats became true carnivores at least 15 million years ago. Generally, animals whose diet contains a wide range of foods are better adapted to changing environments than those that rely on a single or limited range of foods. Many highly successful mammals, such as rodents, are omnivorous in their dietary habits and adapt to varying diets and protein levels by adjusting their metabolism. What is the basis for the success of true carnivores, which are extreme specialists, have survived and evolved using a single type of food?

Researchers think the animal tissue consumed provides complete nutrition for the cat and restricted food availability is a trade-off for the high nutritive value of the food. Animals that eat other animals are not faced with deficiencies of essential nutrients, provided there is an abundance of prey. However, a diet of only animal tissue does not provide the balance of nutrients demanded by the cells of the body. However, if strict carnivores, such as the cat, evolved adaptive metabolic pathways to handle this imbalance of nutrients, then these adaptations were a possible key to the success of carnivores. One example of a metabolic pathway adaptation is the cat's ability to produce vitamin C in its liver. The cat doesn't need, in fact it can't use, orange juice as people do to get vitamin C.

Cats don't have to worry about gathering, capturing, and eating fruit or its juices. They don't care where their next orange will come from because they make all the Vitamin C they need inside their own personal chemical factory (the liver).

On the other hand, there are numerous nutrients and chemicals that cats need that they can only acquire if they eat animal-derived tissues. That is, they need to prey on other living creatures that do make the essential chemicals that cats don't. Evolutionarily, the cats that hunted, captured and consumed this prey to "borrow" the prey's nutrients survived and reproduced at a higher rate than those that did not.

Most importantly, cats are **carnivores.** The cat cannot sustain its life unless it consumes meat in some form. There are numerous nutrients and chemicals cats need that they can only acquire if they eat animal-derived tissues. That is, they need to prey on other living creatures that do make the essential chemicals that cats do not. Out of necessity, the cat has evolved ways to hunt down, capture and eat this prey to "borrow" the prey's nutrients. Cats are very efficient hunters and have never had to adapt to a low protein diet.

Chapter Six

When Did People Start Associating with Cats?

According to some hypotheses, scavenging of predator kills may have provided a food source that helped early forms of humans spread out of Africa into Asia and Europe around 1.8 million years ago. The earliest direct evidence of how modern humans viewed members of the cat family comes from cave art (Figure 44). Drawings of lions dating to 32,000 years ago were found in 1994 in a cave in Chauvet,

Figure 44 Cave Art of a Cat

France. Lion images carved from mammoth tusk ivory found in Vogelherd caves (Figure 45) in Germany may be even older. Curiously,

Figure 45 Vogelherd Caves In Germany

no lions shown in ice age art from Europe have manes, suggesting ancient European lions (Figure 46) lacked this feature. These drawings

Figure 46 European Lion

not only tell us that people shared the environs with the cats; it also gives us additional clues as to the cats' appearance, their visible characteristics. Some of these characteristics, such as the lion's mane, may not be visible in a fossilized image.

The most important event in relations between humans and cats occurred around 12,000 to 10,000 years ago when a small wildcat began to associate with early human settlements in the Fertile Crescent region of the Middle

East. As humans adopted agriculture, stored grain and other food products attracted rodents and other pests that became ready prey for the local wildcat *Felis silvestris libyca* (Figure 47).

Figure 47 African Wildcat

In 2004, it was reported that people tamed cats as pets at least 9,500 years ago, when researchers unearthed the grave of a prehistoric tabby in Cyprus. The Stone Age cat appears to have been carefully placed alongside a human corpse, along with offerings including jewellery and stone tools.

Until now, historians thought the ancient Egyptians first domesticated cats about 4000 years ago. However, evidence suggests cats were culturally important outside Egypt long before that. Stone and clay figurines of cats up to 10,000 years old have turned up in Syria, Turkey and Israel.

Over time, some of the wildcats lost their more aggressive wild behavior and adapted to life alongside people. Humans found the cat a useful animal to control vermin and an enjoyable companion as a pet.

The DNA from living cats shows that all modern domestic cats around the world are descendants of five females of this wild Middle Eastern subspecies, according to research published in 2007. Prior to this DNA study, many experts thought that cats were first domesticated in ancient Egypt around 2500 BC.

A different kind of interaction with cats is hunting them. For generations, people have hunted cats to protect livestock or human populations. Cats have also been hunted for sport and for the fur trade. The pelts of spotted cats such as leopards and ocelots have been particularly valued. Body parts from tigers, leopards, and other cats have been used in traditional medicine and in magic. This hunting activity, usually illegal, has brought several species of cat to the brink of extinction.

Chapter Seven

Will the New Cats Change and Become Old Cats?

Probably, because evolution is ongoing. As has been the case all along, what will change depends on several factors. Most of these factors are not under the cats' control. One of these factors is climate change. Climate changes (Figure 48) in past geologic epochs have

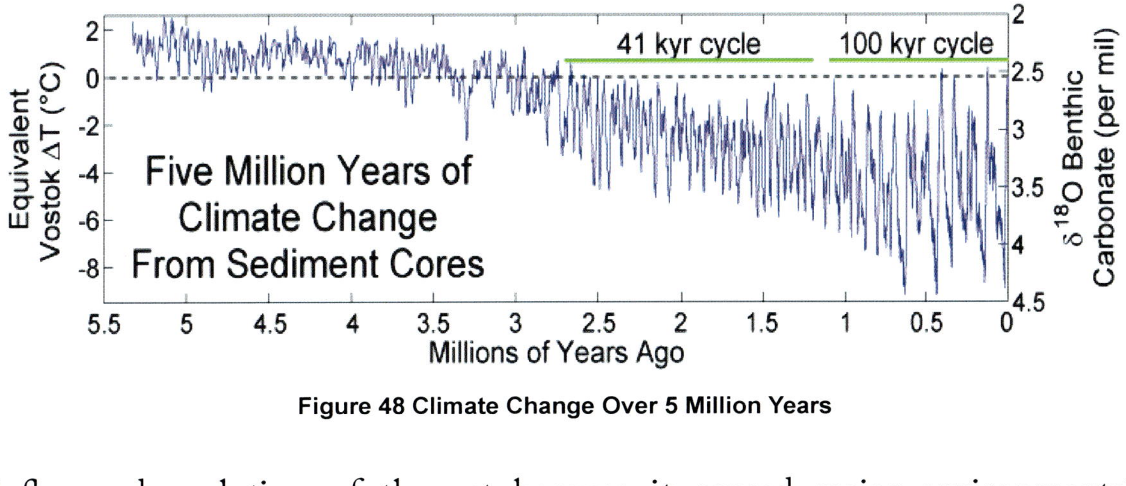

Figure 48 Climate Change Over 5 Million Years

influenced evolution of the cat because it caused major environmental changes, from wet to dry—forest to grassland. These changes altered the cats' available food source. Cats had to either change or die. While many died, some changed.

Wild cats today face the same challenges as their ancestors. The one major factor not in play before is MAN. Humans affect almost all events that challenge survival, even weather, scientists say. Human-caused climate change is a threat to all living organisms. Man's use of fossil fuels has accelerated a previously gradual cycle of weather. During natural weather cycles, plants and animals, though not always successful, were able to change or evolve to meet their new needs. With the weather cycles now being more erratic and, sometimes violent, plants and animals will die before they have a chance to adapt to the new environment. This will result in mass extinctions.

Domesticated cats face a somewhat different challenge. They no longer have to depend on available prey or survival from attacks by larger predators. Humans, who provide for most of their needs, care for most house cats. Nevertheless, these cats too will change. Just how they will change is impossible to predict. It is safe to say that the change will be slow, slower certainly than their wild cousins will. For example, if they no longer have to fight for survival and catch prey, will they lose their claws? Perhaps, but probably not unless having claws makes them unwanted as pets—and only those with short, dull or no claws were kept. Every cat owner would have to agree to this, and, for it to happen naturally, it would still take at least thousands if not millions of years. Having the cat declawed will not affect the cat population because declawing a cat has no genetic effect. It will not be passed on to the next generation. Therefore, you see, there are many problems when trying to predict how evolution will progress. It's a wait and see situation.

To be sure, we have licensed a being that displays its 'otherness' and flaunts its resistance to human interests. This is part of the common view of cats: we value their independence. From time to time, they might want us, but they don't need us. Dogs, by contrast, are said to be fawning and needy, always eager to please. Dogs confirm us; cats confound us and in ways that delight us.

"Because a cat can hunt without eating but cannot eat without hunting, hunting means life to a cat."

Glossary

Adaptation A change in a species that improves its chances of survival.

Biology The study of living things.

Biostratigraphy The part of paleontology that is directly related to the conditions and order of depositions of sedimentary rocks.

Browser An animal that eats young shoots and foliage.

Calamites Ancient plants that were capable of achieving heights of around 30 feet. They grew even as sediments filled in around them, burying them alive.

Carnivora A group of meat eating animals that are genetically related

Carnivore An animal that feeds on meat.

Chalcitheres (kal-si-theres') Large plains type animal resembling a cross between a horse and gorilla, which inhabited Southern Europe, Asia, and Africa during the Miocene.

Chromosome A package for carrying DNA in the cells.

Common Ancestor The most recent ancestral form or species from which two different species evolved.

Cordaites Primitive trees that went extinct millions of years ago. No lineage of descendants survives today.

DNA A complicated organic acid that carries genetic information in the chromosomes. Each molecule of DNA consists of millions of genes, responsible for characteristics, such as hair and eye color.

Eastern Beringia Unglaciated Alaska, Yukon and adjacent Northwest Territories.

Epoch A period marked by unusual or important events.

Erosion The natural process of wearing away by water or wind.

Evolution The process by which populations and species change over time.

Fauna The animal life typical of a region, period, or special environment.

Felidae The cat family.

Fertile Crescent A crescent-shaped region in the Middle East incorporating the Levant, Ancient Mesopotamia, and **Ancient Egypt**, known as the "Cradle of Civilization." The region was named the "Fertile Crescent" because of its rich soil and half-moon shape.

Fossil A trace or print or the remains of a plant or animal of a past age preserved in earth or rock.

Gene A unit of DNA that controls the development of a single characteristic in an individual.

Genera Plural of genus.

Genetics – a branch of biology that deals with the heredity and variation of living things.

Genus Closely related organisms but of different species.

Grazers Animals that eat plants.

Herbivorous Eating or living on plants.

Hybrid An animal or plant whose parents differ in some hereditary characteristic or belong to different breeds or species.

Hyoid bone A U-shaped bone at the base of the tongue that supports the tongue muscles.

Hypothesis An idea or explanation not proved but assumed to be true for purposes of argument or further study.

Lepidodendron Known as the "scale tree" because of the distinctive diamond-shaped leaf scars that covered its outer bark midsection.

Metabolic pathways A series of **chemical** reactions occurring within a **cell**.

Mutation An event that changes the sequence of the DNA in a gene.

Natural Selection The process by which favorable **heritable traits** become more common in successive **generations** of a population of reproducing **organisms**, and unfavorable heritable traits become less common.

Nimravidae Sometimes known as false saber-tooths, an extinct family of mammalian carnivores. Although some Nimravids physically resembled the saber-toothed cats of genus *Smilodon*, they were not closely related, but evolved a similar form through parallel evolution.

Nocturnal Active at night.

Omnivorous Eating both plants and animals as food.

Paleontology The science of the forms of life existing in former geological periods as represented by fossil animals and plants.

Sediment - Material such as stones and sand carried onto land or in the water by wind, water or a glacier.

Sigillaria An extinct tree with leaves and roots very similar to its cousin, the **lepidodendron**, but it differed in that it had much less branching and its tall, columnar trunk lacked the scale-pattern of lepidodendron, instead exhibiting straight, fluted furrows along the trunk. Occasionally, the trunks were smooth.

Species A group of organisms capable of interbreeding and producing fertile offspring.

Stratigraphy A branch of geology that studies and interprets beds or layers of earth and rock.

Theory A well-substantiated explanation of some aspect of the natural world that typically incorporates many confirmed observations, laws, and successfully verified **hypotheses**.

Variation The differences you see among individuals.

Facts of Evolution

Natural selection is the cornerstone of Darwin's theory of evolution.

Genetic drift, along with *natural selection*, are the driving forces changing the frequency of existing genes through time.

Adaptations are the traits that allow organisms to survive the selection process.

"Gene flow" between the different populations of a species keeps a species coherent over time.

A combination of natural selection and gene flow between its members serves as a force that maintains the integrity of a species.

Chance is an often-overlooked occurrence that has affected greatly the evolutionary direction of species.

All members of a population look like one another, small *variations* exist from one to another.

Variation is a key requirement for evolution.

Offspring of all living things tend to *inherit* the traits of their parents.

For *evolution* to occur, *variation* must be *heritable*.

Evolution is **the** key scientific principle behind every substantive thing we know about biology.

Mutations are changes in an organism's DNA.

Though they occur all of the time, only certain *mutations* are important to *evolution*: those that are *heritable.*

Mutations don't occur in response to environmental factors; they already exist in a population. In the presence of a new environmental factor that makes the *mutation* beneficial, the organism with that *mutation* is more likely to survive and reproduce.

Speciation is the process where a single lineage splits into two lineages. New species arise from existing species.

Appendix I
Status and Population Trends for Endangered Cats

Common name	Range	1996	2008	Trend
African Golden	Central Africa	Least Concern	Near Threatened	Declining
Andean Cat	South America	Vulnerable	Endangered	Declining
Asiatic Golden	Southeast Asia	Near Threatened	Near Threatened	Declining
Black-footed	Southern Africa	Least Concern	Vulnerable	Declining
Bobcat	North America	Least Concern	Least Concern	Stable
Bay Cat	Borneo	Vulnerable	Endangered	Declining
Canada Lynx	North America	Least Concern	Least Concern	Stable
Caracal	Africa, Asia	Least Concern	Least Concern	Unknown
Cheetah	Africa, Asia	Vulnerable	Vulnerable	Declining
Clouded Leopard	Asia	Vulnerable	Vulnerable	Declining
Eurasian Lynx	Eurasia	Least Concern	Least Concern	Stable
Fishing Cat	Asia	Near Threatened	Endangered	Declining
Flat-headed Cat	Asia	Vulnerable	Endangered	Declining
Geoffroy's Cat	South America	Least Concern	Near Threatened	Declining
Iberian Lynx	Spain	Endangered	Critically Endangered	Declining
Jaguar	Americas	Near Threatened	Near Threatened	Declining
Jaguarundi	Americas	Least Concern	Least Concern	Declining
Jungle Cat	Asia, Africa	Least Concern	Least Concern	Declining
Guina (Kodkod)	South America	Vulnerable	Vulnerable	Declining
Leopard	Africa, Asia	Least Concern	Near Threatened	Declining
Leopard Cat	Asia	Least Concern	Least Concern	Stable
Lion	Africa, India	Vulnerable	Vulnerable	Declining
Manul (Pallas')	Asia	Least Concern	Near Threatened	Declining
Marbled Cat	Asia	Data Deficient	Vulnerable	Declining
Margay	Americas	Least Concern	Near Threatened	Declining
Ocelot	Americas	Least Concern	Least Concern	Declining
Oncilla	Americas	Near Threatened	Vulnerable	Declining
Pampas Cat	South America	Least Concern	Near Threatened	Declining
Puma (Cougar)	Americas	Least Concern	Least Concern	Declining
Rusty-spotted Cat	India	Data Deficient	Vulnerable	Declining
Sand Cat	Asia, Africa	Least Concern	Near Threatened	Unknown
Serval	Africa	Least Concern	Least Concern	Stable
Snow Leopard	Asia	Endangered	Endangered	Declining
Tiger	Asia	Endangered	Endangered	Declining
Wildcat	Africa, Asia, Europe	Least Concern	Least Concern	Declining

http://www.wildcatconservation.org/wild-cats/status-of-wild-cats/

Appendix II
Wild Cat List by Continent

#	Common Name	Species	Continent
1	Andean cat	*Leopardus jacobita*	South America
2	Pampas cat	*Leopardus colocolo*	South America
3	Guina, Kodkod	*Leopardus guigna*	South America
4	Geoffroy's cat	*Leopardus geoffroyi*	South America
5	Oncilla, Tigrina	*Leopardus tigrinus*	South America
6	Margay	*Leopardus wiedi*	South America
7	Ocelot	*Leopardus pardalis*	South America
8	Jaguarundi	*Puma yaguarondi*	South America
9	Jaguar	*Panthera onca*	South America
10	Puma	*Puma concolor*	South, North America
11	Bobcat	*Lynx rufus*	North America
12	Canadian lynx	*Lynx canadensis*	North America
13	Eurasian lynx	*Lynx lynx*	Europe, Asia
14	Iberian lynx	*Lynx pardina*	Europe
15	Black-footed cat	*Felis nigripes*	Africa
16	African golden cat	*Profelis aurata*	Africa
17	Serval	*Leptailurus serval*	Africa

18	Cheetah	*Acinonyx jubatus*	Africa
19	Lion	*Panthera Leo*	Africa
20	Caracal	*Caracal caracal*	Africa, Asia
21	Leopard	*Panthera pardus*	Africa, Asia
22	Jungle cat	*Felis chaus*	Asia
23	Sand cat	*Felis margarita*	Africa, Asia
24	Wildcat	*Felis silvestris*	Asia, Africa, Europe
25	Chinese mountain cat	*Felis silvestris ssp. bieti*	Asia
26	Pallas's cat	*Otocolobus manul*	Asia
27	Snow leopard	*Panthera uncia*	Asia
28	Tiger	*Panthera tigris*	Asia
29	Leopard cat	*Prionailurus bengalensis*	Asia
30	Fishing cat	*Prionailurus viverrinus*	Asia
31	Rusty-spotted cat	*Prionailurus rubiginosus*	Asia
32	Asian golden cat	*Pardofelis temminckii*	Asia
33	Marbled cat	*Pardofelis marmorata*	Asia
34	Clouded leopard	*Neofelis nebulosa*	Asia
35	Sunda Clouded leopard	*Neofelis diardi*	Asia
36	Bornean bay cat	*Pardofelis badia*	Asia
37	Flat-headed cat	*Prionailurus planiceps*	Asia

Species Reference: IUCN 2012. IUCN Red List of Threatened Species. Version 2012.2.
www.iucnredlist.org
www.wildcatfamily.com/wild-cat-lists
April 2013

More Books about Cats

Ames, Lee J. 2012. Draw 50 Cats: The Step-by-Step Way to Draw Domestic Breeds, Wild Cats, Cuddly Kittens, and Famous Felines. Paperback – Amazon

Beck, Paul. 2011. Uncover a Cat (Uncover Books). Hardcover – Amazon

Bunnell, Kenny. 2013. Cats: Cat Puzzle Book--Interactive and Fun! Kindle Edition

Carney, Elizabeth. 2011. National Geographic Kids Everything Big Cats: Pictures to Purr About and Info to Make You Roar! Paperback – Amazon

Charlie the Cavalier. 2014. Cat Jokes for Kids: Hilarious Jokes (Best Clean Joke Books for Kids). Kindle Edition.

Clutton-Brock, Juliet. 2014. DK Eyewitness Books: Cat. Paperback – Amazon

DeSpain, Pleasant. 2003. Tales of Cats (Books of Nine Lives Book 9). Kindle Edition

Ford, Kevin. 2015. CATS: Wild Cats (Animals For Kids Book 1). Kindle Edition

McDonald, Barry J. 2013. Cats (Amazing Pictures and Fun Facts Book About Cats). Kindle Edition

Montgomery, Sy. 2014. Chasing Cheetahs: The Race to Save Africa's Fastest Cat (Scientists in the Field Series) Hardcover – Amazon

Petras, Kathryn and Ross Petras. 1998. Fandex Family Field Guides: Cats. Paperback – Amazon.

Walker, Anne. 2014. Cats: 80 Fascinating Facts For Kids About Cats. Kindle Edition

Illustration Credits

https://en.wikipedia.org/wiki/Wildcat

“”Wiki-Felis sylvestris” by Zoologist - Own work. Licensed under CC BY-SA 3.0 via Wikimedia Commons - https://commons.wikimedia.org/wiki/File:Wiki-Felis_sylvestris.png#/media/File:Wiki-Felis_sylvestris.png

*What is a Sabertooth? *Smilodon*, the saber toothed tiger. University of California, Museum of Paleontology

http://prehistoric-fauna.com/Proailurus-lemanensis

https://en.wikipedia.org/wiki/Felidae

http:// www.creationhistory.com/freisenhahncave.shtml

http://www.ibtimes.co.uk/cave-paintings-drawings-hunter-gatherers-brazil-photos-520670

http://www.ice-age-art.de/anfaenge_der_kunst/vogelherd.php

http://www.planetseed.com/relatedarticle/temperature-change-history

http://willemsvdmerwe.deviantart.com/art/Protoceratids-401841991

https://simple.wikipedia.org/wiki/Nimravidae

O’Brien, Stephen J. and Warren E. Johnson. 2007. Genetics: The evolution of cats. Scientific American 297(1):68-75.

Shorter, C. M. 2006. Evolution of Wild Cats - Genetic Time Line. http://Tigerhomes.org

http://www.tarpits.org The Page Museum is located at the Rancho La Brea Tar Pits in the heart of Los Angeles.

Resources

Christiansen, Per. 2007. Comparative bite forces and canine bending strength in feline and sabretooth felids: implications for predatory ecology. Zoological Journal of the Linnean Society 151(2): 423-437.

Currier, Mary Jean P. 1983. *Felis concolor.* The American Society of Mammologists, Mammalian Species No.200:1-7.

Foss, Scott E. 1995. Fossils of the Badlands of South Dakota. Fact Sheet, Badlands Natural History Association, Badlands National Park, SD.

Futuyma, Douglas J. 1986. Evolutionary Biology. Sinauer Associates , Sunderland, MA.

Hayes, Dann. KU Researcher Labels New Saber-tooth Fossil Find Fantastic. dhayes@ ukans.edu.

Hemmer, Helmut. 1972. *Unica unica*. The American Society of Mammalogists, Mammalian Species 20:1-5.

http://www.tarpits.org The Page Museum is located at the Rancho La Brea Tar Pits in the heart of Los Angeles.

http://www.concordia.edu/html/faculty/meissner/fries_home. Welcome to Friesenhahn Cave.

http:// https://en.wikipedia.org/wiki/Felidae

https://en.wikipedia.org/wiki/Wildcat. Domestication

http:// www.pantherkut.com/big

http://www.pbs.org/wgbh/evolution/library/05/2/l_052_02.html

http://academic.reed.edu/biology/courses/BIO342/2011_syllabus/2011_websites/ Purring%20CH&JP/index.html

http:// http://icatcare.org/

http:// www.creationhistory.com/freisenhahncave.shtml

https://www.newscientist.com/.../dn13307-darwin-still-controversial-after-all- these-years/

Kitchener, Andrew. 1991. The natural history of the wild cats. Comstock Publishing Associates, Ithaca, NY.

Kürschner, Wolfram M., Zlatko Kvaček, and David L. Dilcher. 2008. The impact of Miocene atmospheric carbon dioxide fluctuations on climate and the evolution of terrestrial ecosystems. Proceedings of the National Academy of Sciences 105(2):449-453.

Legrand-Defretin, Veronique. 1994. Differences between cats and dogs: a nutritional view. Proceedings of the Nutrition Society 53:15-24

Martin, L.D., J.P. Babiarz, V. L. Naples and J. Hearst. 2000. Three Ways To Be a Saber-Toothed Cat. Naturwissenschaften 87:41–44

Mazák, Vratislav. 1981. *Panthera tigis*. The American Society of Mammalogists, Mammalian Species 152:1-8.

Muir, Hazel. 2004. Ancient remains could be oldest pet cat. Daily news 8 April 2004, New Scientist https://www.newscientist.com/article/dn4867-ancient-remains-could-be-oldest-pet-cat/

Murray, Julie L. and Gregory L. Gardner. 1997. *Leopardus pardalis*. The American Society of Mammalogists, Mammalian Species 548:1-10.

O'Brien, Stephen J. and Warren E. Johnson. 2007. Genetics: The evolution of cats. Scientific American 297(1):68-75.

Seymour, Kevin L. 1989. *Panthera onca*. The American Society of Mammalogists, Mammalian Species 340:1-9.

Turner, Alan. 1997. The big cats and their fossil relatives. Columbia University Press, New York, NY.

Wastluber, J. History of the domestic cat and cat breeds. Chapter1 *In*: www.vetmed.ucdavis.edu/ccah/local-assets/pdfs/FelHusCh1.pdf

Weiss, Rick. 2005. Cats' Sweet Tooth Long Gone; Genetic Mutation Leaves Felines Unable to Taste Sugar. *Washington Post*, 25 July 2005, A, p. 6.

Made in United States
North Haven, CT
05 November 2026

21765350R60058